LEVEL **2**

Arctic Animals

Jennifer Szymanski

NATIONAL GEOGRAPHIC

Washington, D.C.

To the staff of public libraries everywhere— thank you for the quiet space. —J.S.

Published by National Geographic Partners, LLC, Washington, DC 20036.

Designed by Anne LeongSon

The author and publisher gratefully acknowledge the expert content review of this book by Jeff W. Higdon, wildlife biologist, Higdon Wildlife Consulting, Manitoba, Canada, and the literacy review of this book by Mariam Jean Dreher, professor emerita of reading education, University of Maryland, College Park, and fact-checking by Michelle Harris.

Library of Congress Cataloging-in-Publication Data
Names: Szymanski, Jennifer, author.
Title: Arctic animals / Jennifer Szymanski.
Description: Washington, D.C. : National Geographic, 2023. | Series: National geographic readers | Audience: Ages 5-8 | Audience: Grades 2-3
Identifiers: LCCN 2021019716 (print) | LCCN 2021019717 (ebook) | ISBN 9781426339936 (paperback) | ISBN 9781426339943 (library binding) | ISBN 9781426339950 (ebook) | ISBN 9781426339967 (ebook other)
Subjects: LCSH: Animals--Arctic regions--Juvenile literature. | Arctic regions--Juvenile literature.
Classification: LCC QL105 .S99 2023 (print) | LCC QL105 (ebook) | DDC 591.998--dc23
LC record available at https://lccn.loc .gov/2021019716
LC ebook record available at https://lccn.loc .gov/2021019717

Photo Credits
Cover, jimcumming88/Adobe Stock; 1, Paul Nicklen/National Geographic Image Collection; 3, Eric Isselee/Shutterstock; 4-5, James.Pintar/ Shutterstock; 4-29 (UP), BigMouse/Shutterstock; 4, Ronan Donovan/National Geographic Image Collection; 6-32, Wirestock/Adobe Stock; 7, jack stephens/Alamy Stock Photo; 7-29, Vect0r0vich/ Getty Images; 8-9, Andy Trowbridge/Nature Picture Library; 10, Donald M. Jones/Minden Pictures; 11 (LO), jamenpercy/Adobe Stock; 11 (UP), Ronan Donovan/National Geographic Image Collection; 12, Dgwildlife/Getty Images; 13, Patrick J. Endres/Getty Images; 14-15, Ralph Lee Hopkins/National Geographic Image Collection; 15-32, Wirestock/Adobe Stock; 16, Paul Nicklen/National Geographic Image Collection; 17 (UP), WaterFrame/Alamy Stock Photo; 17 (LO), Paul Nicklen/National Geographic Image Collection; 18 (UP), RLS Photo/Adobe Stock; 18 (LO), Norbert Wu/Minden Pictures; 19, Scenics & Science/Alamy Stock Photo; 20, ZSSD/ Minden Pictures; 21-32, Orsolya Haarberg/ National Geographic Image Collection; 22-23, littleting/Pradthana Jarusriboonchai/Getty Images; 23 (UP RT), neurobite/Adobe Stock; 24 (UP), Vladimir Melnik/Adobe Stock; 24 (LO LE), Andrey Nekrasov/Getty Images; 24 (LO RT), Paul Loewen/Shutterstock; 25 (CTR LE), Paul Souders/ Getty Images; 25 (UP RT), Pav-Pro Photography/ Adobe Stock; 25 (LO LE), Paul Nicklen/National Geographic Image Collection; 26-32, Geoffrey Reynaud/Getty Images; 26, Mark Medcalf/Getty Images; 28-32, Ingo Arndt/Nature Picture Library; 29, Brian Skerry/National Geographic Image Collection; 30 (LO LE), Paul Nicklen/ National Geographic Image Collection; 30 (UP RT), bublik_polina/Adobe Stock; 31 (UP RT), nirutft/Adobe Stock; 31 (LO), Josef Svoboda/ Adobe Stock

Printed in the United States of America
22/WOR/1

Contents

A Wintry World

A fox runs across a snowy plain. Seals dive into an icy ocean. A walrus sleeps on a rocky shore. Owls hoot and wolves howl.

This is the Arctic.

An arctic wolf may howl when it finds food or thinks danger is near.

Unlike most owls, the snowy owl hunts during the day.

Much of the Arctic is covered by water and ice.

Temperatures here are often below freezing. There is little rain. Many animals cannot live here.

Yet some animals are able to survive in the Arctic. They can find food, stay warm, and stay safe. Let's meet some of them!

Arctic hares tuck in their paws and sit to keep warm.

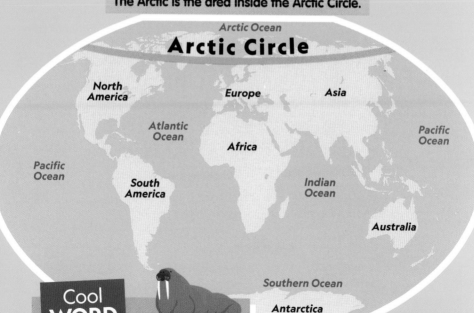

The Arctic is the area inside the Arctic Circle.

Arctic Ocean

Arctic Circle

North America

Europe

Asia

Atlantic Ocean

Pacific Ocean

Africa

Pacific Ocean

South America

Indian Ocean

Australia

Southern Ocean

Antarctica

Cool **WORD**

SURVIVE: To stay alive

Life on Land

Most of the year, the land is covered by snow and ice. Animals that look white blend into the background.

This arctic fox is hard to see when it's hunting. It can sneak up on other animals. Arctic foxes are not choosy predators. They eat anything they can find.

Cool WORD

PREDATOR: An animal that hunts and eats other animals

Q Why was the fox hunting rabbits?

A Because it was in the mood for fast food!

Foxes can hear and smell animals that are underneath the snow.

Arctic foxes are not the only meat-eaters in the Arctic. Check out these animals and how they use their bodies to hunt.

Rough-Legged Hawk

Rough-legged hawks fly above the land. Their sharp eyes can spot a meal on the ground below. Then they dive down to grab it.

Arctic Wolf

Wolves chase hares and other speedy prey. A wolf's wide, furry paws help to keep it from sliding as it runs.

Wolverine

Wolverines can smell rodents such as squirrels and lemmings underground. They dig through ice and dirt with strong claws to eat them.

Not all Arctic animals eat meat. Plant-eaters live in the Arctic, too.

Musk oxen kick the snow with their hooves to find the plants below. Strong teeth help them chew tough plants.

Musk oxen have thick, shaggy coats to keep them warm.

A ptarmigan can use its sharp beak to snip off twigs to eat.

Plant-eaters are often hunted by other animals. To help protect themselves, many Arctic animals can blend in with the snow and ice. When the ptarmigan (TAR-mih-gen) fluffs its white feathers, it looks like a lump of snow. Foxes and wolves may not see it.

Life in the Sea

The Arctic Ocean is the smallest ocean on Earth. In winter, it is almost completely covered by ice. Still, many animals live in its waters.

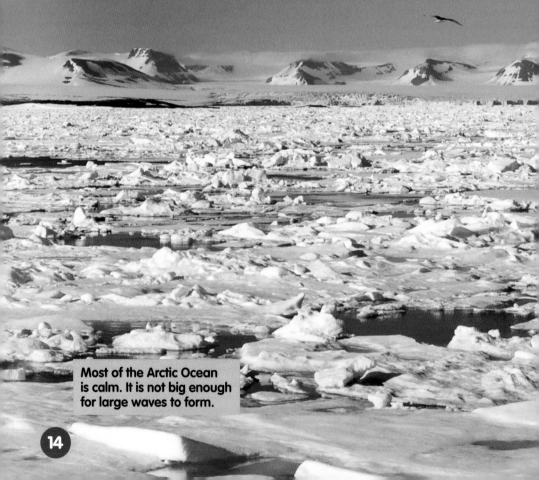

Most of the Arctic Ocean is calm. It is not big enough for large waves to form.

Whales are the largest ocean animals. A thick layer of blubber covers their bodies under their skin. This keeps them warm in the icy water. Some whales, like humpbacks, visit the Arctic Ocean only in summer.

When a whale breaks through the surface of the water, it is called breaching.

Cool **WORD**

BLUBBER: A thick layer of fat that some animals have to keep them warm

Other kinds of whales live in the Arctic Ocean all year. They find plenty of food to eat.

Narwhal

A narwhal's body glides easily through water. It can dive deep and find fish at the bottom of the ocean.

Beluga

A beluga whale uses sound to find food. It makes a sound and listens for the echo. The sound bounces off fish, telling the whale where to find them. The fish will make a great meal.

Bowhead

A bowhead's mouth is very big. It scoops a lot of tiny animals out of the water at one time.

Skates can hide their flat bodies by covering them with sand and mud.

Jellyfish aren't fish! They are Cnidarians (Nye-DARE-ee-inz).

Smaller animals swim in the icy water, too. Schools of fish are food for sharks. Skates glide near the bottom of the ocean. Jellyfish float near the top. There are even some kinds of life that are too small to see without special tools!

Zooplankton are tiny animals that live in water. This photo was taken under a microscope. The zooplankton look much, much bigger here.

Life on the Coast

A polar bear can stay underwater for more than two minutes before it needs to take a breath of air.

Some animals live in the ocean and on land. Polar bears hunt on sea ice and swim in the ocean. When the ice melts, polar bears must go on land. They are one of the world's largest predators.

In the ocean, walruses can swim fast. Their strong back flippers push them through the water.

A walrus rests often, but sometimes it will not sleep for many days in a row.

They rest on the ice and on land.

21

Many kinds of birds soar high over the Arctic coast. Some birds, like puffins, nest on rocky cliffs.

Puffins live in groups called colonies.

Puffins are
good hunters.
They can swim
and fly very fast.
Their beaks can hold
a lot of fish at once.

6 AWESOME Arctic Animal FACTS

1

Walruses use **their tusks** to help **pull** themselves **out of the water** and **onto the sea ice.**

2

Beluga whales can **swim** upside down and **backward.**

3

Arctic hares have **big back feet.** They help hares **walk on snow without sinking—** just like a person using snowshoes!

Polar bears are excellent **swimmers,** thanks to their **webbed feet.**

4

5

Both male and female **caribou** have **antlers.** Males **lose** their antlers in **winter.** Females usually don't **lose** theirs until spring. Then **both** the males and females **grow** their antlers **back again.**

6

Male narwhals have a **superlong tusk** that is actually a **tooth!** This tusk is the reason for the narwhal's **nickname, "unicorn of the sea."**

More Ways to Survive

Many animals survive the cold Arctic winter in other ways.

Animals may migrate from the Arctic to other places. Some large herds of caribou move to areas with more plants to eat. They rest and have babies. Then they come back to the Arctic.

Some caribou herds migrate to warmer areas when snow starts to fall.

Both mother and father terns care for their chicks and bring them food.

Terns fly halfway around the world! In fall, they fly from the Arctic to Antarctica. Then they return in the spring to have chicks.

A ground squirrel spends winters in a warm and cozy hole.

Other animals hibernate in winter. There is not much food. So they rest for many months.

A ground squirrel's body gets cooler. Its heart beats slowly. It curls up in its nest until spring.

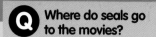

Q Where do seals go to the movies?

A The dive-in!

Animals who live in the Arctic fit perfectly into their cold home!

Harp seals spend most of their time in the Arctic's icy water.

Cool **WORD**

HIBERNATE: To spend the cold winter months resting. An animal's body slows down when it hibernates.

QUIZ WHIZ

How much do you know about life in the Arctic? After reading this book, probably a lot! Take this quiz and find out.

Answers are at the bottom of page 31.

1 The area of Earth known as the Arctic is _____.

A. in the rainforest
B. in Australia
C. inside the Arctic Circle
D. where penguins live

2 Which of these animals uses sound to find food?

A. walrus
B. polar bear
C. arctic tern
D. beluga whale

3 Terns and caribou are animals that migrate. They _____.

A. climb trees
B. dive deep into the ocean
C. sleep all winter
D. move from one area to another

4 Which two words best describe the Arctic?

A. cold and snowy
B. warm and wet
C. hot and cold
D. rainy and sunny

5 Some animals have a thick layer of fat called _____.

A. puffins
B. blubber
C. antlers
D. fur

6 Which Arctic animal is one of the world's largest predators?

A. musk ox
B. arctic hare
C. puffin
D. polar bear

7 When animals _____, they rest and use less energy.

A. eat
B. hibernate
C. migrate
D. communicate

GLOSSARY

BLUBBER: A thick layer of fat that some animals have to keep them warm

HERD: A large group of animals that live together

HIBERNATE: To spend the cold winter months resting. An animal's body slows down when it hibernates.

MIGRATE: To move from one area to another to find food or a mate or to have babies

PREDATOR: An animal that hunts and eats other animals

SURVIVE: To stay alive